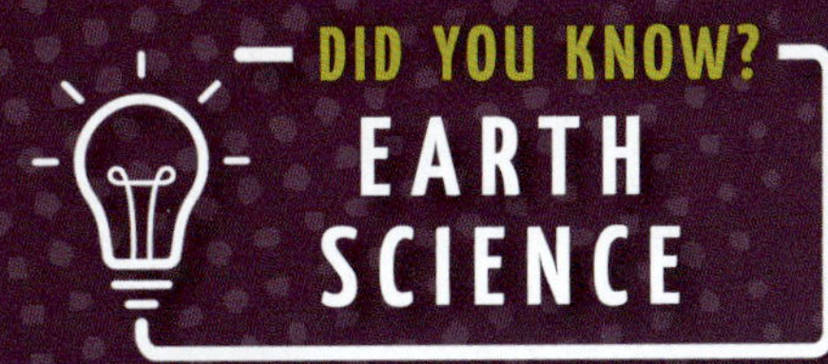

20 THINGS YOU DIDN'T KNOW ABOUT

ASTRONOMY

MARIE MORRISON

PowerKiDS press

Published in 2023 by The Rosen Publishing Group, Inc.
2544 Clinton Street, Buffalo, NY 14224

Portions of this work were originally authored by Jill Keppeler and published as *20 Fun Facts About Astronomy*. All new material in this edition was authored by Marie Morrison.

Editor: Greg Roza
Book Design: Tanya Dellaccio

Photo Credits: Cover Zakharchuk/Shutterstock.com; p 5 Mykola Mazuryk/Shutterstock.com; p. 6 NASA, JPL-Caltech, Susan Stolovy; p. 7 NASA; p. 8 Dima Zel/Shutterstock.com; p. 9 Teemu Tretjakov/Shutterstock.com; p. 10 Aphelleon/Shutterstock.com; p. 11 Solar Orbiter/EUI Team (ESA & NASA); pp. 12, 14, 19 NASA images/Shutterstock.com; p. 13 NASA/Johns Hopkins University Applied Physics Laboratory/Southwest Research Institute; p. 15 Triff/Shutterstock.com; p. 16 SSV, MIPL, Magellan Team, NASA; p. 17 Peyker/Shutterstock.com; p. 18 AlexLMX/Shutterstock.com; p. 20 pongpinun traisrisilp/Shutterstock.com; p. 22 thipjang/Shutterstock.com; p. 23 Danita Delimont/Shutterstock.com; p. 24 Elzbieta Krzysztof/Shutterstock.com; p. 25 NASA/ESA/HEIC and The Hubble Heritage Team; p. 26 Heritage Image Partnership Ltd/Alamy Images; p. 27 Simon Serdar/Alamy Images; p. 29 vchal/Shutterstock.com.

Cataloging-in-Publication Data
Names: Morrison, Marie.
Title: 20 Things You Didn't Know About Astronomy / Marie Morrison.
Description: New York : PowerKids Press, 2023. | Series: Did You Know? Earth Science| Includes glossary and index.
Identifiers: ISBN 9781538389607 (pbk.) | ISBN 9781538389621 (library bound) | ISBN 9781538389638 (ebook)
Subjects: LCSH: Astronomy–Juvenile literature. | Planets–Juvenile literature. | Space–Juvenile literature.
Classification: LCC QB46 M67 2023 | DDC 520–dc23

Manufactured in the United States of America

CPSIA Compliance Information: Batch #CWPK23. For Further Information contact Rosen Publishing at 1-800-237-9932.

CONTENTS

WHAT'S OUT THERE? 4
STAR LIGHT, STAR BRIGHT 6
SO MANY STARS 8
SUPER SUN 10
PLANETARY POINTS 12
GOOD NEIGHBORS 18
EXOPLANETS OUT THERE 20
AMAZING ASTRONOMERS 22
SPACE TRAVELERS 26
LOOK TO THE FUTURE 28
GLOSSARY 30
FOR MORE INFORMATION 31
INDEX 32

WHAT'S OUT THERE?

For hundreds of years, people have looked up at the night sky and wondered: What's out there? We can see the sun, the stars, the moon, and sometimes **planets**, but it's hard to understand what's so far away.

Humans have made up stories about what we see in the sky, but we've also learned to use science to find out facts too. People—including scientists called astronomers—are making discoveries about objects in space almost every day.

Astronomy is the study of planets, stars, and other things in space—such as Earth's moon, shown here!

STAR LIGHT, STAR BRIGHT

DID YOU KNOW?

Starlight can take thousands of years to reach Earth!

When you look up at the stars, you're really looking back in time. You're seeing their light from many years ago. Light from the farthest stars in the Milky Way **galaxy** can take 100,000 years (or more) to reach Earth.

The Milky Way is the galaxy that includes our sun and **solar system.**

Earth's sun is a yellow star. Blue stars are hotter than yellow ones. However, red stars are cooler, even though they're still very hot: about 3,000 degrees Kelvin (4,940 degrees Fahrenheit, 2,727 degrees Celsius). There are also white stars.

SO MANY STARS

DID YOU KNOW?

There are about 100 billion stars in our galaxy.

Scientists have to **estimate** the number of stars. It's impossible to know the exact number, but they use tools such as **telescopes** to make a good guess. There may be many more—space is big!

The James Webb space telescope launched in December 2021. Scientists will use it to learn many things about outer space.

Using the Hubble space telescope, scientists have estimated there are about 2 trillion galaxies. That means there could be about 200 billion trillion stars, a huge number. That's a "2" with 23 zeroes after it!

SUPER SUN

DID YOU KNOW?

The sun contains almost all of the mass in our solar system!

Our sun is small compared to many other stars—but it contains 99.8 to 99.9 percent of our entire solar system's mass. Jupiter has most of the rest of the mass. Earth only has a very tiny part.

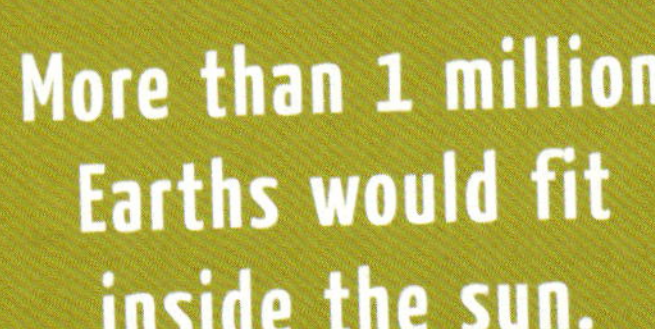

More than 1 million Earths would fit inside the sun.

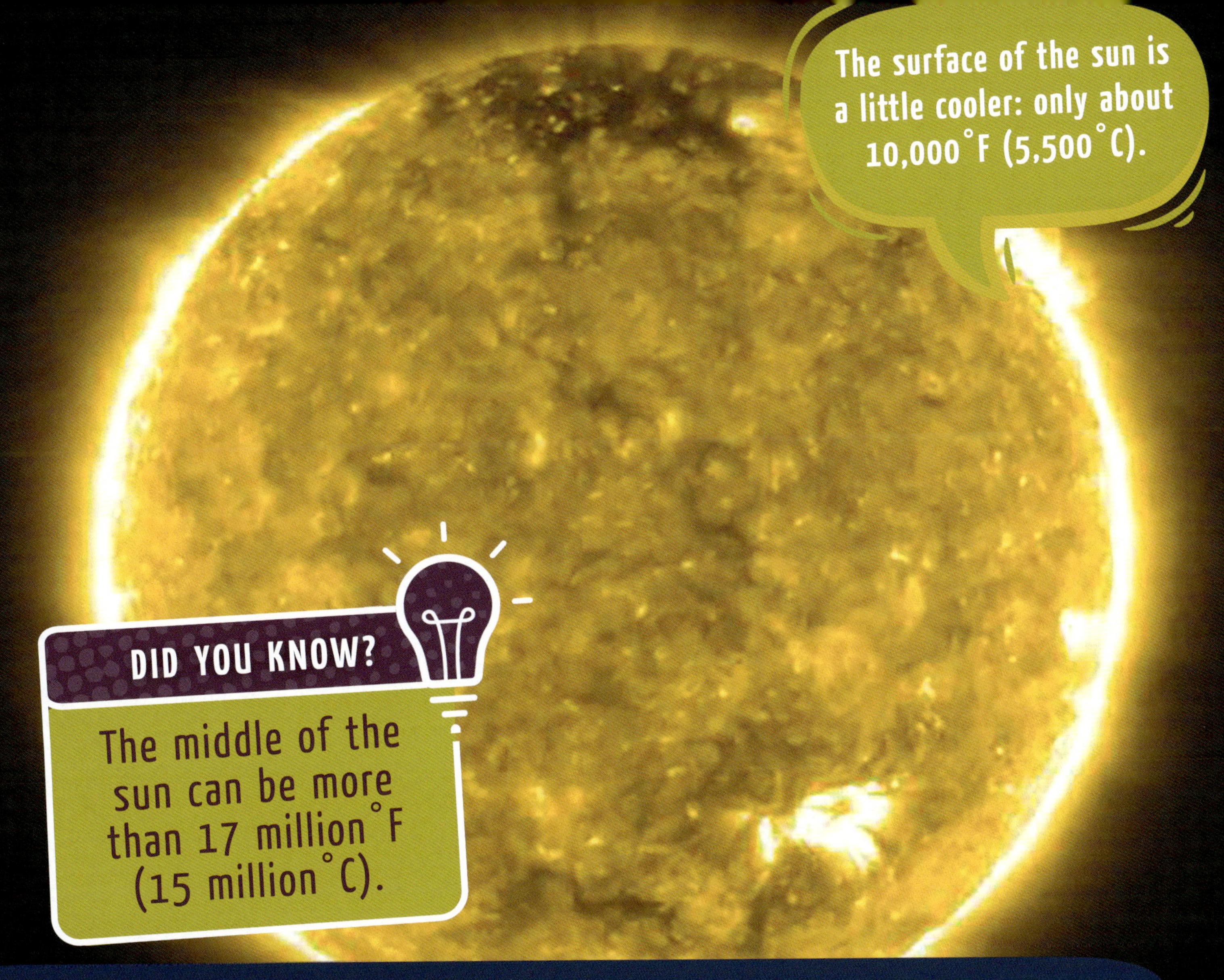

NASA says it would take 100 billion tons (90.8 billion MT) of **dynamite** exploding every second to match the sun's energy. That's a lot of big booms!

PLANETARY POINTS

DID YOU KNOW?

The former planet Pluto is smaller than Earth's moon.

The moon is about 2,200 miles (3,540 km) wide. Pluto is about 1,400 miles (2,250 km) wide. It was considered a full planet until 2006, when scientists started calling it a dwarf, or small, planet instead.

Pluto is named for the Greek or Roman god of the underworld. This image from NASA shows what it might look like.

Pluto is very cold. It can reach -387°F (-233°C)!

DID YOU KNOW?

Pluto may have active ice **volcanoes!**

In March 2022, scientists studying photos of Pluto found signs of ice volcanoes. These volcanoes may still be active—which means there could be liquid water (or something like it) under the surface.

DID YOU KNOW?

The planet Uranus rolls over and over instead of spinning round and round.

This planet, a gas giant, looks like it rolls around the sun like a ball! This means it has very strange seasons. During two seasons of each year, half of the planet never gets sunlight at all.

A year on Uranus is 84 Earth years, so a single season is about 21 Earth years long.

Our solar system includes the planets Mercury, Venus, Earth, Mars, Jupiter, Saturn, Uranus, and Neptune.

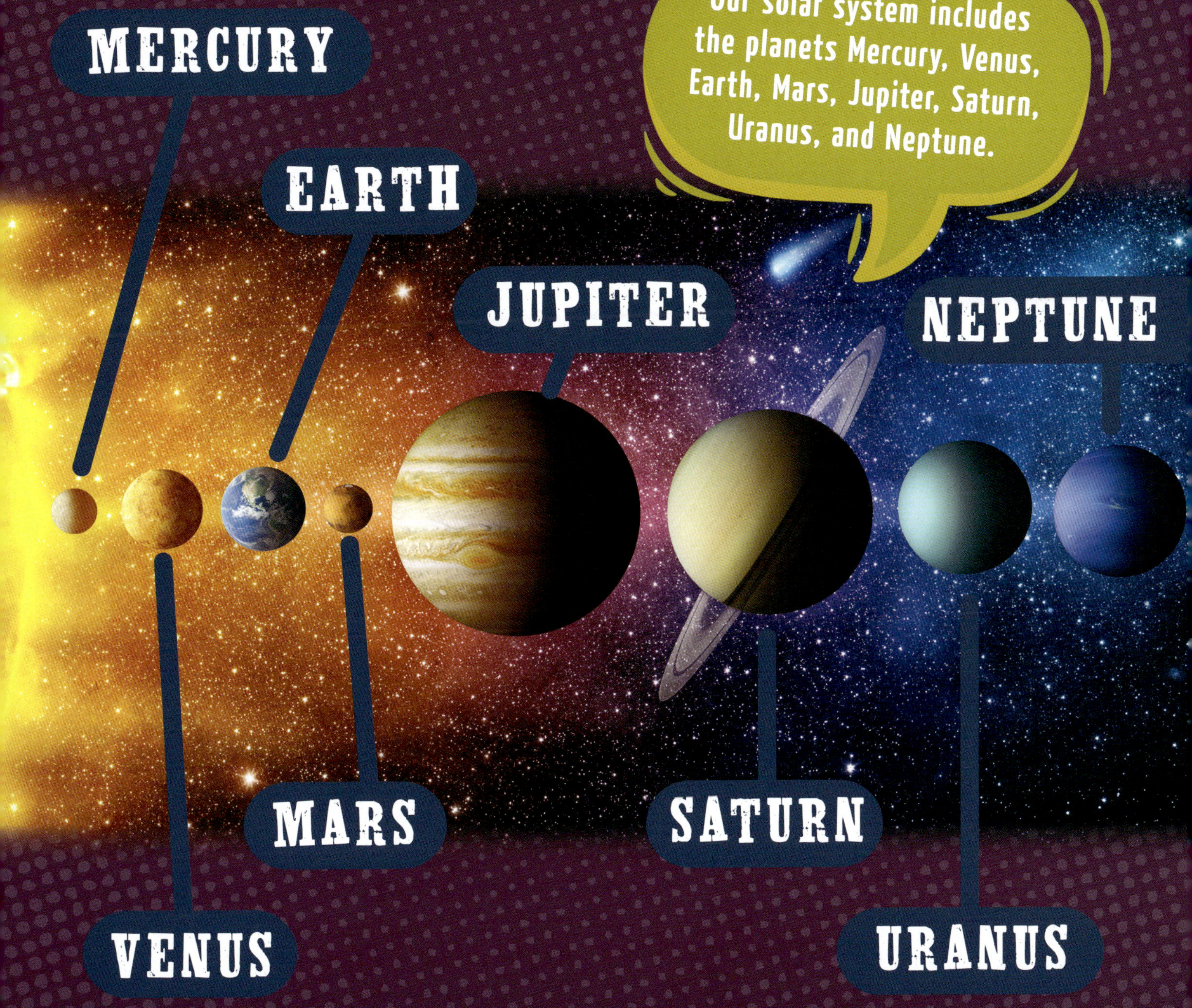

DID YOU KNOW?

On Venus, the sun rises in the west.

On Earth, if you're awake to watch a sunrise, you'll see the sun appear slowly in the east. On Venus, it's the opposite! This is because Venus spins clockwise, or to the right. The rest of the planets in our solar system spin to the left.

Scientists believe Venus flipped over at some point.

What would you guess is the hottest planet in our solar system? Mercury is the closest planet to the sun. However, Venus is the hottest planet! Its surface temperature can be nearly 900°F (482°C).

GOOD NEIGHBORS

DID YOU KNOW?

On average, Mercury is every planet's nearest neighbor!

How is this possible? Mercury **orbits** the sun more closely than any other planet. So, its nearest and farthest distances from the other planets aren't that different. Other planets spend much more time on opposite sides of the sun from each other.

Venus is the planet that gets the closest to Earth at one time—but Mercury is the closest on average.

The Goldilocks zone (named because it's "just right") of a star is the range in which planets might have (or once have had) liquid water. However, neither Venus nor Mars have liquid water now.

EXOPLANETS OUT THERE

DID YOU KNOW?

Astronomers have confirmed more than 5,000 planets outside our solar system.

Exoplanets are planets that orbit a star that isn't our sun. As of March 2022, NASA has confirmed, or made sure of, the discovery of more than 5,000 of them. However, we know this is only a tiny fraction of the exoplanets out there.

The first exoplanet was discovered by two astronomers in 1992.

THE EXOPLANET FILES

The closest exoplanet discovered so far is about 4.2 **light-years** away. It's orbiting Proxima Centauri, the closest star to our sun.

The farthest away exoplanet to be discovered so far is about 17,000 light-years away.

Some exoplanets aren't in orbit around a star. They're called rogue exoplanets and drift through space on their own.

Scientists have found at least one exoplanet that's made partly of the **mineral** diamond.

The largest exoplanet discovered so far is about nine times the size of Jupiter.

Some of the discovered exoplanets seem to be in their own suns' Goldilocks zones. So far, scientists have found more than 50.

Scientists have found a few planets outside our solar system that orbit two suns.

Exoplanet information is changing all the time! In fact, even the above facts might change any day as scientists discover new planets and new information.

AMAZING ASTRONOMERS

DID YOU KNOW?

There were prehistoric astronomers.

Some researchers say cave paintings made as many as 40,000 years ago in Spain, France, Turkey, and Germany may show star constellations. Ancient carvings may also show signs that these long-ago peoples studied the stars.

A constellation is a group of stars that forms a shape and has been given a name. We can't be sure these paintings show constellations, but researchers think they might!

An observatory is a building or place where people observe natural phenomena, or things. Many civilizations created ways to track the sun, stars, moon, and planets throughout the sky. Some are buildings, but some are outdoor sites.

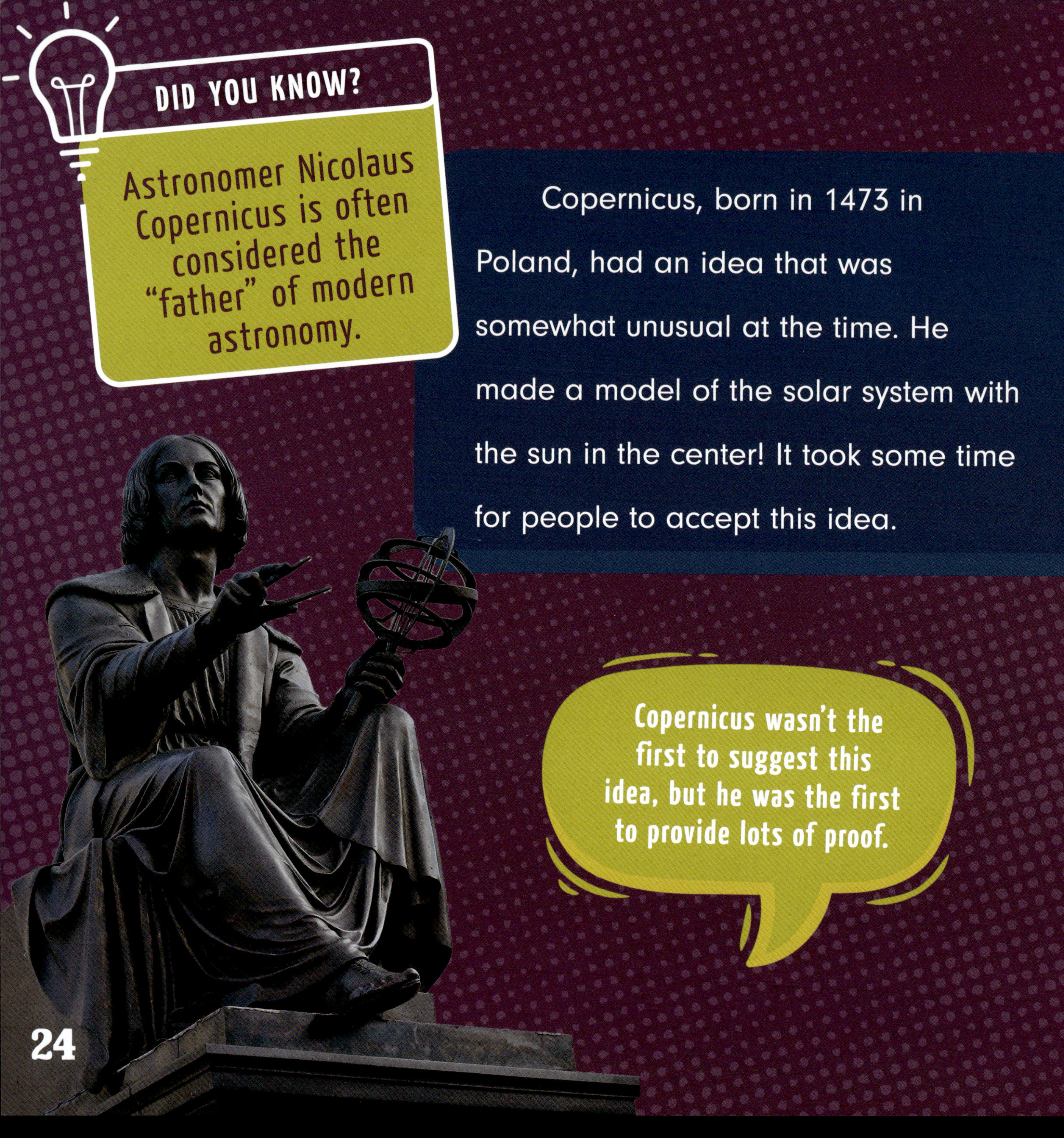

Copernicus, born in 1473 in Poland, had an idea that was somewhat unusual at the time. He made a model of the solar system with the sun in the center! It took some time for people to accept this idea.

DID YOU KNOW?

Kids have made some amazing astronomy discoveries!

In 2011, Kathryn Gray of Canada was only 10 when she was studying pictures of space taken by her family. She noticed a spot of light in one but not another. It was a supernova! Two years later, her brother, Nathan, discovered one too.

SPACE TRAVELERS

DID YOU KNOW?

Two dogs (and some smaller animals) were the first living things to orbit Earth and return safely.

Dogs Belka and Strelka went to space in August 1960 on the Soviet *Sputnik 5* spacecraft. They returned the next day. The dogs traveled with a rabbit, some mice, and some rats, which also returned safely.

One of Strelka's puppies was later given to U.S. President John F. Kennedy in 1961. Her name was Pushinka.

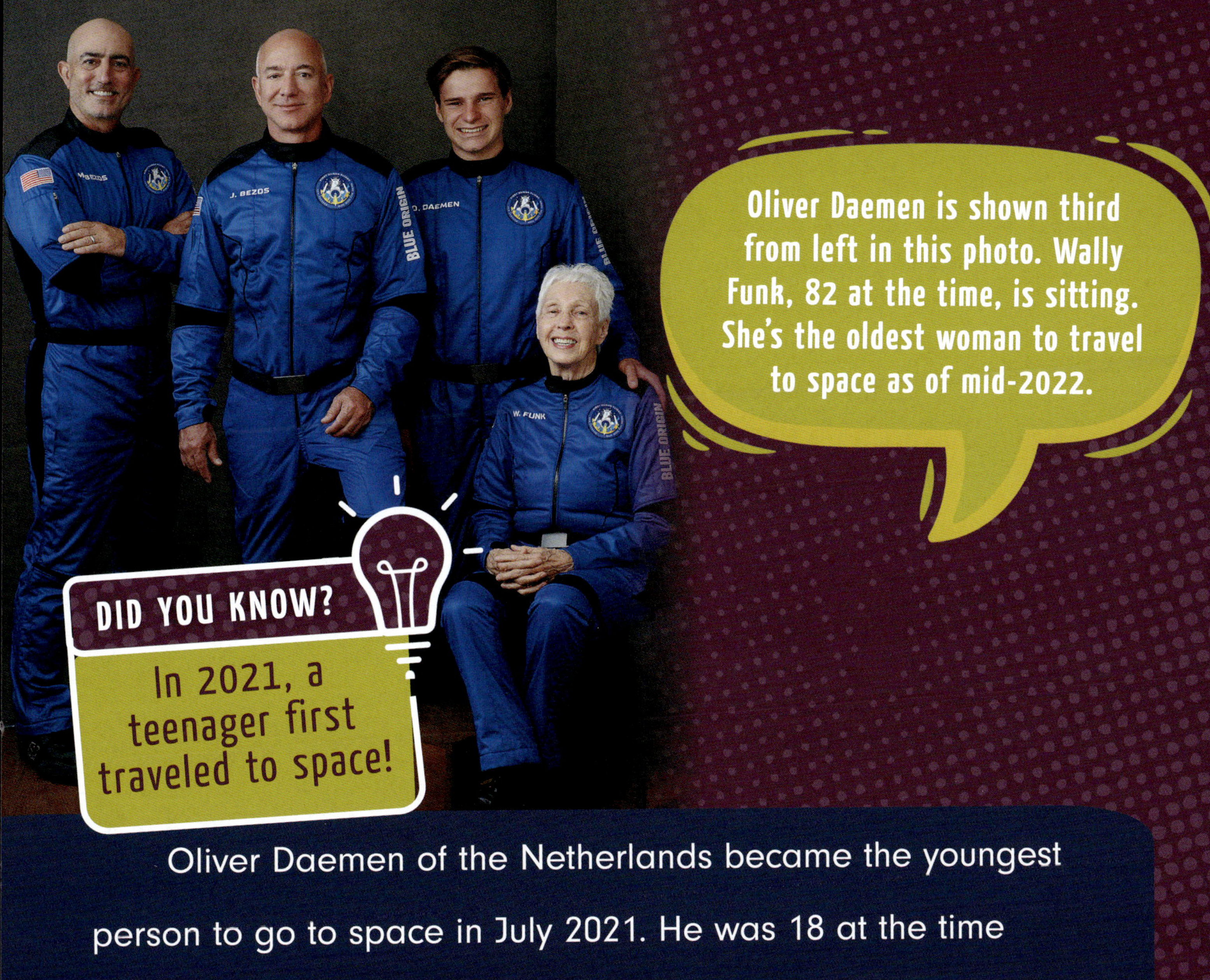

Oliver Daemen of the Netherlands became the youngest person to go to space in July 2021. He was 18 at the time and had just graduated from high school. He took part in an 11-minute trip on a rocket created by the company Blue Origin.

LOOK TO THE FUTURE

What we know about astronomy and space is changing every day. Scientists keep making new discoveries and inventing new tools to help us learn more about our neighbors (near and far) in space.

Countries and companies are planning new missions to space as well. In fact, hundreds of launches were set for 2022, and there will likely be even more in 2023. The James Webb space telescope will be sending back its first images. It's an exciting time to learn more about astronomy!

What exciting new discoveries in astronomy would you like to see?

GLOSSARY

dynamite: A powerful explosive often found in the form of a stick.

estimate: To make an educated guess about the size or nature of something.

galaxy: A large group of stars, planets, gas, and dust that form a unit within the universe.

light-year: The distance light can travel in one year.

mineral: Matter in the ground that forms rocks.

orbit: To travel in a circle or oval around something, or the path used to make that trip.

planet: A large, round object in space that travels around a star.

solar system: The sun and all the space objects that orbit it, including the planets and their moons.

telescope: A tool that makes faraway objects look bigger and closer.

temperature: How hot or cold something is.

volcano: Usually, an opening in a planet's surface through which hot, liquid rock sometimes flows. Sometimes used for an opening in a planet's surface through which something else pours out.

FOR MORE INFORMATION

BOOKS

Becker, Helaine. *National Geographic Kids: Everything Space.* Washington, D.C.: National Geographic, 2018.

Betts, Bruce. *Astronomy for Kids: How to Explore Outer Space with Binoculars, a Telescope, or Just Your Eyes.* Emeryville, CA: Rockridge Press, 2021.

WEBSITES

NASA Science SpacePlace
spaceplace.nasa.gov/
This website by NASA offers many space-related activities and facts for kids who are interested in astronomy.

Passport to Space
kids.nationalgeographic.com/space
National Geographic offers a wealth of information about planets and other topics.

Publisher's note to educators and parents: Our editors have carefully reviewed these websites to ensure that they are suitable for students. Many websites change frequently, however, and we cannot guarantee that a site's future contents will continue to meet our high standards of quality and educational value. Be advised that students should be closely supervised whenever they access the internet.

INDEX

C

Copernicus, Nicolaus, 24

G

Goldilocks zone, 19, 21

J

Jupiter, 10, 15, 21

M

Mars, 15, 19

Mercury, 15, 17, 18, 19

Milky Way, 6

N

Neptune, 15

P

Pluto, 12, 13

Proxima Centauri, 21

S

Saturn, 15

Sputnik 5, 26

T

telescopes, 8, 9, 28

U

Uranus, 14, 15

V

Venus, 15, 16, 17, 18, 19